Magnetotherapy
and Natural Healing

Wayne Bottiger, M.Ed.

ISBN: 978-1-4243-3171-0

Table of Contents

Introduction

In medicine there are a number of various types and styles used in the treatment of illness and disease. Medical professionals around the world provide a measure of services to individuals who seek their help and assistance in some way. Western based medicine has become one of the largest businesses in the world and continues to grow. This is why most people, for reasons of finance, have found the assistance of medical professionals beyond their reach.

As the health care system continues to grow there have been a number of different approaches developed for the treatment and management of illness and disease. These systems have been labeled as *alternative medicine*. While the word alternative suggests something outside of the norm or something less effective, the truth is these alternative approaches to healing have begun to provide more options for people who are being turned away from western medicine.

From the many different types of natural medicine there have arisen specific treatments and remedies for illness and disease at a fraction of the costs normally charged for western interventions. One such treatment method is called *Magnetotherapy*. This book will examine the underlying principles of Magnetotherapy, and provide an educational background for the study and practice of this holistic approach to healing.

It should be understood that the information in this book is presented for education and research purposes only. The methods and descriptions are presented so that readers will have a better understanding of Magnetotherapy, and should not be considered a prescription for use by individuals seeking medical health. Anyone seeking medical attention should always consult with a physician before using this or any other method of healing.

Chapter 1 – Magnetotherapy Defined

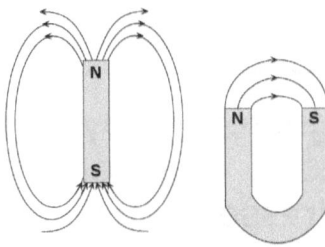

"Our ancestors derived more from the all-wise nature than from the artificial salts and synthetic drugs."

M.T. Santwani

This chapter will examine the definition and meaning of Magnetotherapy. Additionally, it will provide a look at the arguments for and against this therapeutic approach from variety of individuals who cite a numerous research studies in support of their arguments. As a point of interest it is important to understand there are many possibilities for using methods which are considered unconventional by allopathic medicine. Perhaps the best way to define something is by looking at it from the perspective of those who oppose it, as well as those who support its use, in order to have a complete view of its meaning and purpose.

There are many ways to describe the approach of Magnetotherapy. First of all, it is a system many attribute to trace to Paracelsus (1493-1543). Paracelsus was a physician who studied alchemy extensively, and he reasoned that magnetic fields existed in the body which provided channels through which illness and disease could be removed with the use of magnets (Livingston, 1998).

When considering the meaning and purpose of Magnetotherapy one specific question must be addressed: *How plausible is this approach in terms of medicine?* Barrett (2002), contends there is 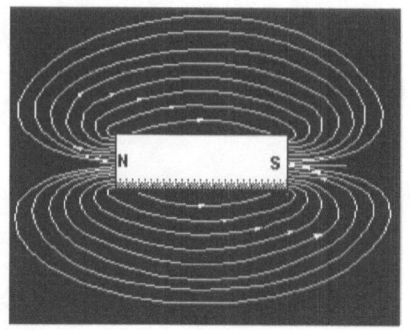 no scientific basis for the use of magnets in the treatment of illness and disease, citing the fact that many of the products available on the market produce no significant magnetic effects on the body. Ramey (1998), points out that magnetic therapy is nothing more than a pseudoscience that offers no significant effects on body tissues, and as such should not be considered a plausible approach to healing.

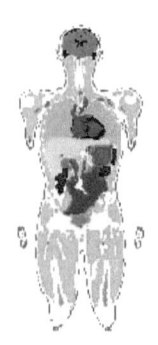 McCrory (2005), labels the use of magnets for therapeutic purposes to be nothing more than *hocus science*. He cites no connection between the use of magnets and healing in studies on patients experiencing a variety of ailments. His conclusion is that the use of magnetic therapy works simply as a *placebo* (any application or substance used as a medicine or treatment while having no pharmacological effect) (Dictionary.com, *Placebo* 2007).

 In a letter written to William L. Roper, a manufacturer of tectonic magnets sold as a therapeutic device, Gill (1997) cites a number of apparent problems with magnets being advertised as a capable method of producing healing effects for various ailments in the body. Her letter was intended to serve as a cease and desist order against Roper, citing a number of Federal Food and Drug Administration rules disapproving the use of such devices.

Carroll (2006), calls Magnetotherapy a "type of alternative medicine" which attracts individuals whom he labels as *gullible* in regards to their search for specific cures of illness and disease. Carroll points to numerous research studies supporting his claims against the use of magnets for healing purposes.

With so much negative attention being given regarding the use of magnetic devices for the treatment of illness and disease it is easy to understand why this particular approach is labeled quackery or folk medicine. However, readers should examine all of the evidence before making a determination of its usefulness. Consider the following points: (1) a great number of reports point to some connection between the use of magnets and healing, even though they are seemingly unverifiable; (2) scientific studies regarding the use of magnets for healing purposes are often conducted in controlled environments which favor those opposed to the use of such devices; and (3) those opposed to the use of magnets for therapeutic purposes often have some vested interest in providing a negative view of this system of treatment.

From a holistic point of view, Magnetotherapy is defined as: (1) a healing art which uses the application of magnets to the body for the purpose of removing the source and symptoms of disease (IBAM, *Magnetotherapy,* c. 2000); (2) a healing system using magnets to effect the healing of living tissues and organs in the body (Santwani, 2005); (3) a system utilizing magnetic fields to affect the body's internal systems and organs, which result in the healing of health related conditions (IndianGyan.com, 2000); and (4) a therapy applying magnets to different parts of the body to create a magnetic field which effects the healing of specific health related conditions (MedicineNet.com, *Definition of Magnet Therapy*, 2007).

The first responsibility of those working in the healing arts is to the people they provide a service to. It is unlikely that the debate over the effectiveness or ineffectiveness of magnets used for healing will end any time soon. It appears that both sides of the argument have a vested interest in supporting their claims. This is why articles and research studies are necessary to validate the claims of those in favor of magnetic therapies. From the stand point of education it is important to provide an accurate and unbiased resource so readers and students can formulate their own opinion based on the facts.

Chapter 2 – The Theory of Magnetotherapy

"That Magnetic healing is nothing new can be seen by looking at early records of scientifically advanced civilizations."
Gary Null

To understand the theory of Magnetotherapy it is important to provide a background and history of magnetic theory. There are several relevant theories which will help explain some of the underlying premises for Magnetotherapy. By examining the current theories on magnetic fields it will be easier for readers to understand the actions taking place when Magnetotherapy is used. This chapter is divided into three parts: (1) The History of Magnetic Theory; (2) The Principle Theory of Magnetotherapy; and (3) The Relationship between Magnetic Theory and Magnetotherapy.

The science of magnetism is much broader than the scope of this chapter. Therefore, readers are encouraged to conduct their own research into the various different theories of magnetic fields. Without question the subject is worth the investigation, especially in terms of understanding the principles of Magnetotherapy. The study of magnetism has resulted in advances in science and medicine over several thousand years. It is for this reason that the inclusion of information regarding magnetic theory is important as part of the learning continuum focused on Magnetotherapy.

The History of Magnetic theory

Timeline

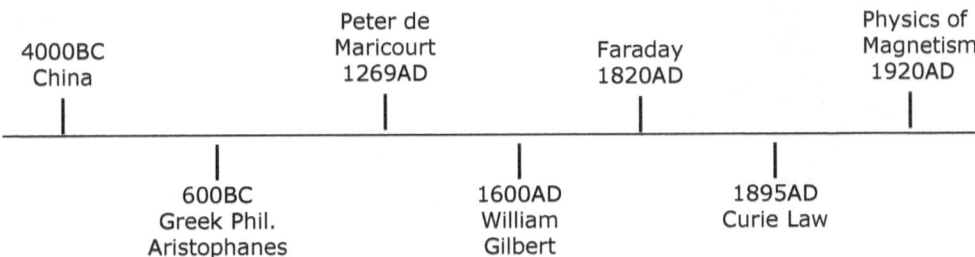

| 4000BC China | | Peter de Maricourt 1269AD | | Faraday 1820AD | | Physics of Magnetism 1920AD |
| 600BC Greek Phil. Aristophanes | | | 1600AD William Gilbert | | 1895AD Curie Law | |

The complete history of magnetism contains too many references to include in one chapter. A great number of scientists and researchers contributed to the study of magnetism and as such, they opened the door to natural science that changed the way mankind sees the universe. The periods on the timeline above outline a brief look at the significant events takings place in the area of magnetism.

Early magnetism dates back approximately 4,000 years to ancient China. Chinese practitioners were believed to use crude compasses which had needles showing the relative location of the north 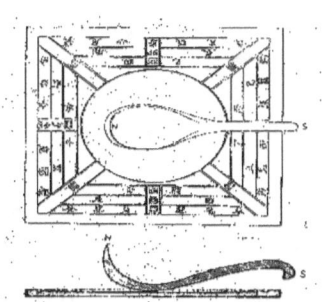 and south poles. Their inventions may have been connected to Arabic or Indian origins. However, much is unknown regarding specific experiments and tests which were conducted during this period of history.

Early modern magnetism can be traced back to experiments using lodestones (magnetite rock formations containing magnetic properties, capable of attracting iron or steel), which were described by Greek philosophers. Early experiments surrounding magnetism were conducted by a philosopher named Aristophanes who took amber (a yellow translucent mineral), and rubbed it against fur, causing it to attract materials such as bird feathers and light fabrics (Lee, 1970).

It was also during this time that the great philosophers of the period set forth their beliefs concerning magnetism. Unfortunately, they simply tried to fit observations from their experiments into an academic model rather

than explaining them in terms of science and nature. Two such philosophers were Thales (636-546 BC) – he supported the idea that lodestones possessed a life of their own; and Aristotle (384-322 BC) – he argued that a relationship between the mover and the object being moved existed, which in turn affected the process. It was this logic and reason that provided a sense of mystery concerning the study of magnetism.

Following a period when few publicized experiments were conducted, the study of magnetism reappeared around the middle 1200s. It was then that Pierre de Maricourt created a spherical device, showing the relative position of a geographical location by using diametrically opposite positions. He is credited with being the first person to identify the earth's north and south pole. He reported his findings in a treaty called, *Epistola de magnete* (Wikipedia.com, *Peter de Maricourt*, 2007). In his treaty Maricourt gave credit to others whom he regarded for their research and study in the area of magnetism.

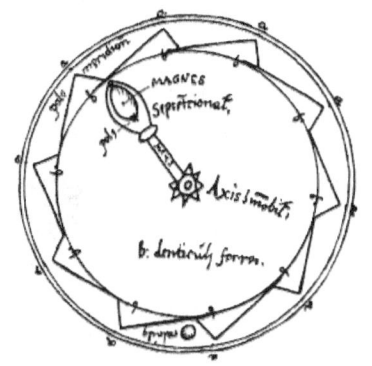

In 1600 AD, Dr. William Gilbert, an English philosopher, put forth the proposition that the earth itself was magnetic, and attempted to explain his theory in a written work called *De Magnete, Magneticisque Corporibus, et de Magno Magnete Tellure* (Wikipedia.com, *William Gilbert*, 2007). Gilbert believed magnetism represented the collective soul of the earth and as such served as an axis point for rotation. He never pointed directly to theoretical assumptions concerning the solar system or universe, but his studies became the cornerstone of research by the physicist Johannes Kepler (Van Helden, 1995).

The significance of Gilbert's theory rested on what he called the *terrella* or <u>little earth</u>. In explaining the earth's magnetism, an object such as a compass would be moved around a magnetized spherical object 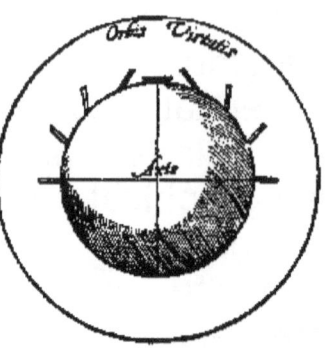 to show that it always pointed north-south. It was Gilbert's further belief that the same activities take place regarding the earth's north and south poles.

 A few hundred years later, a scientist by the name of Michael Faraday (1791 – 1867), an English physicist, was largely responsible for the study and development of the electromagnet. It was in the 1820s when Faraday made his most significant contribution to the field of science and electricity. During this time Faraday began experimenting with electromagnetism. He went to great lengths to demonstrate the process responsible for the transformation of electrical energy into motive force. His studies eventually resulted in the creation of the electric motor.

Some of his more significant experiments included his discovery of electrical induction (1831) and the understanding of inductive capacity (1837). Faraday developed a number of additional theories related to light and gravitational systems (Bellis, 2007).

In 1895 a professor of physics by the name of Pierre Curie, created a theory that emphasized the magnetic properties of different substances at various temperatures. His theory assumed that the magnetism of a given substance would change as the temperature increased or decreased. The point of change was called the *Curie Point* (The Nobel Foundation, *Pierre Curie*, 2007).

Curie was less interested in the politics of science, preferring to spend his time in study and research of different ideas related to natural science. His wife Marie Curie, became more widely known for her discoveries in the field of medicine. It was Pierre Curie's hunger for understanding that led him to make significant discoveries that were later applauded by the scientific community.

From this point, the introduction of many other theories and studies led to a great number of discoveries which formed the basis for modern physics and the understanding of magnetism. Modern physicists like Lorentz,

Gauss, Coulomb, and Ohm made a significant impact in the area of electricity and magnetism (Nave, 2005). Without question, their research discoveries changed the entire view of science regarding the subject of magnetism.

The Principle Theory of Magnetotherapy

From the standpoint of science and medicine, Magnetotherapy is classified as a physical treatment using methods dating back for hundreds of years (ZES Bros. Inc, 2007). It has only been in the past several decades that this method received notoriety as an approach to healing. Those who use Magnetotherapy consider it to be superior as a practical medicine.

What are the principles of Magnetotherapy? The theory of Magnetotherapy is based on several premises including: (1) the earth possesses lines of force called *fields* in which magnetic energy flows in a north-south direction; (2) the farther away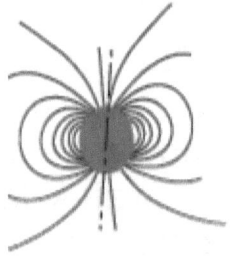

The Earth

from the north or south pole an object is results in a lower value of Gauss (the amount of energy it takes to change the magnetic value of an object in relation to its proximity to the north or south pole); (3) electricity is capable of creating artificial magnetic fields; and (4) low energy magnetic fields are capable of influencing the general properties of human cells, organs and tissues.

How do the principles apply to mankind? Magnetotherapy is capable of creating a magnetic field present in the body so that cells and tissues can auto regulate and repair. The human body is basically made up of cells and tissues through which magnetic energy flows. When a portion of the body or its tissues are expose to magnetism it activates a variety of physiological mechanisms which affect healing.

The Relationship Between Magnetic Theory and Magnetotherapy

From a health perspective Magnetotherapy has important connections to the theory of magnetism. If the premises of magnetic theory are true then it is likely that a viable connection to the practice of Magnetotherapy exists. Magnetic theory states all things existing contain a magnetic field. Magnetotherapy claims that healing is effectuated through the transfer of energy in the body's magnetic pathways.

Today, modern medicine is skeptical of the relationship between the theories of magnetism and Magnetotherapy. However, there are a number of case studies which point to a viable relationship. Additional study and research should be conducted in order to establish a basis of understanding. Once more information becomes available concerning the findings of Magnetotherapy it will be easier for people interested in the applications and procedures of this healing approach to make their own determination. Presently, there is a large number of professionals in the field of medicine who continue to dismiss the application of Magnetotherapy claiming it is nothing more than pseudoscience.

Chapter 3 – The Purpose of Magnetic Induction for Therapeutic Practices

"When properly applied, magnetic therapy enhances the body's natural ability to heal itself rather than simply masking the symptoms."

Robert Becker

To understand the processes in Magnetotherapy it is necessary to look at induction. This chapter will look at the way induction is used for therapeutic purposes in Magnetotherapy. Hopefully, readers will develop a better sense of the entire system of Magnetotherapy so they can begin applying the principles to a healing model.

What is magnetic induction? Simply, induction represents the point at which there is an attraction to an opposite pole (north to south or south to north), creating magnetism (Dictionary.com, *Induction*, 2007) The process of 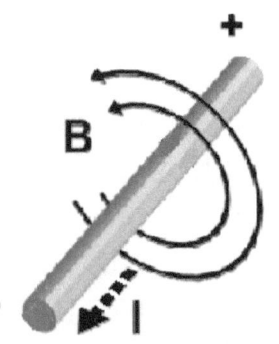 induction is relatively simple to understand. According to the theory of magnetism, things existing possess the ability to carry electrical current. The transfer of current through an object or person is called conduction. The amount of pull or force toward a given object or person is called induction. In the diagram above **B** represents the magnetic field, **I** represents the direction of induction, + and − represent the polarity.

It is the attraction of opposites which creates magnetism. The molecular makeup of a particular object or person determines the amount of its magnetic properties. The magnetic region surrounding an object or person affects the level of induction taking place. When a magnetized object comes within the proximity of an object or person with opposite polarity it begins to create torque (the amount of force necessary to make an object rotate on an axis), which in turn aligns it to a given direction. The proximal size of a magnetic field is called a vector, and represents the area most affected by magnetic force.

It is the magnetic force of an object or person that causes the molecular disruption of equilibrium (the resting state or balance between opposing forces). When a given magnetic field (field of induction)

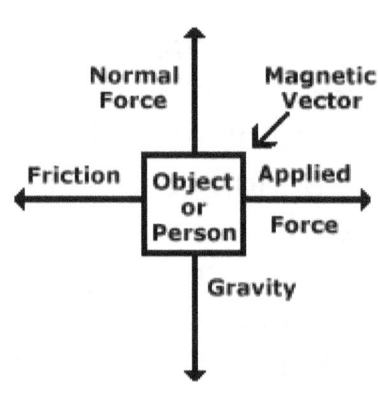

produces a magnetizing force on a vector, it forces charged particles to move in a specific direction towards the field of greatest polarity. Depending on the strength of the magnetic stimulus, current moves from the opposite direction of polarity creating conduction through the field of induction.

Frank Clewer –
(Bunny, 2005)

How does induction affect the human body? The greatest effect on the human body is through a process called *static electricity*. All matter is made up of molecular particles called atoms. Presently there are only one hundred and fifteen different atoms known to man. Each of the different atoms are made up of three parts: (1) nucleus – the central part; (2) protons – electrons with a positive charge; and (3) neutrons – electrons with a negative charge. Normally, the parts of the atom are balanced and in a state of equilibrium. However, the farther away from the nucleus electrons travel affects how tightly they are held. Because the electrons in the outer area of the atom's molecular

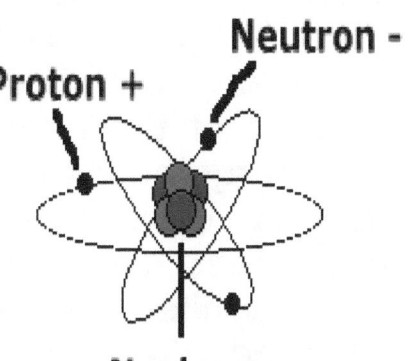

Proton +

Neutron -

Nucleus

structure are loosely held, they tend to move freely between like atoms. When an atom looses electrons and has more protons than neutrons, it is positively charged. When an atom has more neutrons than protons, it is negatively charged. A charged atom is called an *ion*, and is largely responsible for the generation of static electricity (the build up of electricity on an insulated body).

Next, the opposite charges of ions begin to move towards each other when a given force is applied to them. This is also the case when a charged object is attracted to something that is neutral. By charging a given object it is 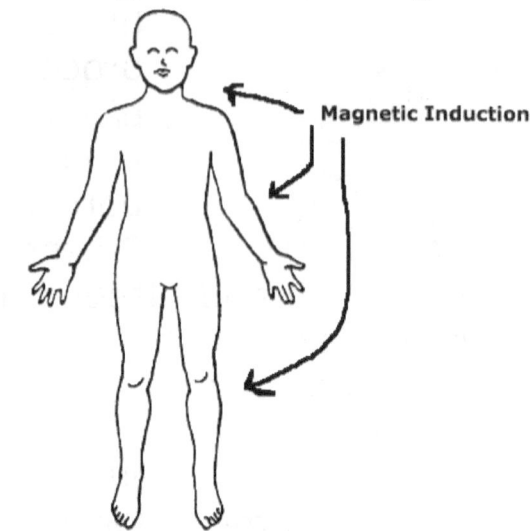 possible to facilitate the flow of electrons to a neutral object. The amount of energy transferred depends on the conductive properties of the neutral object. As a result, a static charge builds up and is transferred to the conductor causing a quick jolt of electrical energy that can contain high amounts of voltage. While the effects of magnetic devices on the body do not produce the same results as static electricity, the principles related to the transfer of energy are the same.

How is induction related to Magnetotherapy? It is the common belief of proponents of magnet therapy that subjecting areas of the body to magnetic devices creates the flow of energy useful in the rejuvenation of cellular tissues. Of course this position has seen its

Magnetic Induction

share of criticism from opponents of Magnetotherapy. The main source of criticism focuses on various scientific claims about magnets being ineffective, and lacking sufficient power to penetrate human skin. Additionally, skeptics point out that there is little evidence to support claims of effectiveness on muscle tissue, bones, blood vessels, and organs cited by holistic practitioners.

A Watchful Conclusion

To date, there are conflicting results in support of both positions held by proponents and opponents of Magnetotherapy as well as the devices used for therapeutic purposes. While the Food and Drug Administration in the United States has not ruled in favor of magnetic devices for the purpose of treating specific health conditions, it has allowed the use of such devices for the treatment of pain so long as the devices make no specific claims of usefulness or effectiveness (SeniorHealth: About.com, 2006). What this means in terms of the use of Magnetotherapy as a whole remains up to the individual. It is likely that additional studies will be done to verify the validity and efficacy of magnetic devices for the treatment of health related conditions.

Chapter 4 – Types and Uses of Therapeutic Magnets

"Magnetism has an intense and far-reaching effect on physical and biological processes of the body."

(IBAM, 1998)

There continues to be great interest in the various types of medical treatment available for the general population. Numerous experiments have shown significant results using methods of healing other than the conventional approaches promoted by western medicine. This chapter will continue to describe the specific medical aspects of Magnetotherapy. As the previous chapter pointed out, there is reason to believe the human body possesses magnetic fields of varying intensity and strength (IBAM, *Guide to Alternative Medicine*, pg. 79, c. 1998).

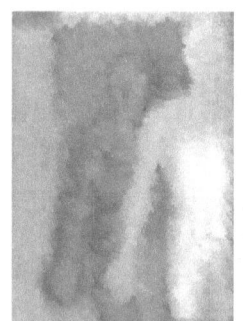

While those supporting the use of Magnetotherapy continue to provide research evidence of their claims, the general sentiment of western medicine remains negative. Some specific questions to be addressed in this chapter include: (1) What types of healing magnets are currently available? (2) What are the general procedures for the use of magnets in terms of a healing model? (3) Therapeutically, what is the optimal period of exposure when magnets are used? (4) What safety implications should be considered when using therapeutic magnets?

What types of healing magnets are currently available? From the standpoint of science, there are several magnets which can be used in the treatment of conditions affecting the body. Because every magnet 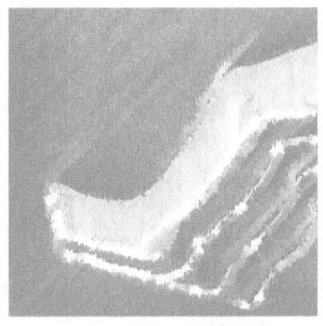 has its own unique properties, it is important to understand the specific ways a magnet interacts with the body's tissues and organs.

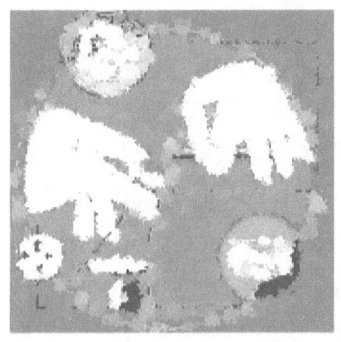 The following properties exist in all magnets: (a) Magnets, regardless of use have a north and south polarity affecting the direction of energy; (b) The polarity of a magnet creates attracting and repelling action; (c) All magnets have a specific field of magnetism which affects their ability to repel or attract energy; and (d) Magnets vary in strength and intensity based on their relative size and use. The magnetic fields of magnets is infinite, and plays a significant role in the therapeutic treatment of specific conditions (IBAM, *Magnetotherapy*, 2000).

The two common artificially manufactured magnets used in Magnetotherapy are (1)

permanent magnets – devices made from composite materials that retain their magnetism once they have been charged. These magnets are commonly used therapeutically by individuals independently; and (2) electro magnets – devices that generate magnetism using electric current. These magnets are commonly used by physicians and therapists in the treatment of medical conditions.

What are the general procedures for the use of magnets in terms of a healing model? Magnetic devices should only be used when they are indicated in the 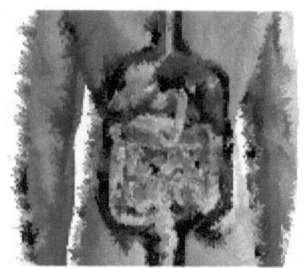 treatment of illness and disease. The following procedural considerations should serve as a guide for use when considering  Magnetotherapy: (1) Determine the specific condition for which magnet therapy is indicated; (2) Select the most appropriate device for treatment, i.e. permanent or electro-magnetic; (3) Determine the specific polarity to be used in treatment; (4) Determine the area of the body that will be exposed to magnet therapy; and (5) Determine the length and number of times magnetic devices will be used in treatment (EBESCO, 2007).

The considerations mentioned is not inclusive of all health related concerns, and as such readers are strongly encouraged to work with professionals trained in the use and application of Magnetotherapies. It is likely that a therapeutic regime can be set up for a number of conditions affecting health. However, it is important to understand the specific indications of any device before using it for health related purposes.

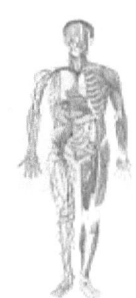 *Therapeutically, what is the optimal period of exposure when magnets are used?* For the most part, magnetic devices are safe when used in accordance with their specific usage. There are some contraindications that professionals and users should be aware of. They include: (1) Placing devices on the body where they will not interfere with the normal function of internal organs that have electric currents flowing through them, i.e. the heart and the brain (Santwani, pg. 66, 2005);

(2) Prolonged exposure to high level magnetic fields such as those generated by MRI machines; (3) Exposure to any level of magnetic influence when other medical devices are implanted into the body such as a heart defibrillator or pacemaker.  It is extremely dangerous to have treatment on areas of the body that could be affected by magnetic currents, disrupting the normal function of these devices; and (4) Other

physiological conditions affecting the brain such as epilepsy. This is only a short list of possible contraindications, and individuals interested in investigating the use of Magnetotherapy are advised to consult a trained professional before beginning any treatment.

What safety implications should be considered when using therapeutic magnets? The number one item among the list of things to consider when using therapeutic magnets is education. As is the  case with any medically related approach or application, individuals should always exercise extreme caution and research specific procedures before using a given method or a type of healing device. Common concerns for safety include: (1) use by individuals with implanted devices such as pacemakers or insulin pumps or defibrillators; (2) women who are pregnant; and (3) individuals with serious medical conditions.

There are a large number of Internet websites available providing a comprehensive list of specific precautions for using magnetic devices for healing purposes. Readers should spend some time researching various aspects of Magnetotherapy. In most cases, the use of magnets has been found safe and free of harmful side effects.

Chapter 5 – The Integration of Magnetotherapy with Other Therapeutic Treatments

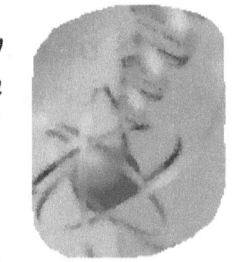

"Various biological rhythms are closely associated with the universal magnetism without which the life would be impossible"
Santwani, M.T.

The integration of Magnetotherapy with other methods of treatment is a subject of much discussion in terms of a modern medical system. There are a number of skeptics and critics in allopathic and natural medicine that argue the effectiveness of Magnetotherapy. From an educational point of view, it is important to understand the processes of Magnetotherapy and their effects on the body in order to determine whether or not it is viable for use with other treatment systems.

This chapter will examine four ideas related to Magnetotherapy and how well it can be used with other therapeutic systems. The four ideas include: (1) The use of Magnetotherapy as a complimentary approach to treatment; (2) The use of Magnetotherapy for the treatment of pain; (3) The interaction of Magnetotherapy when used

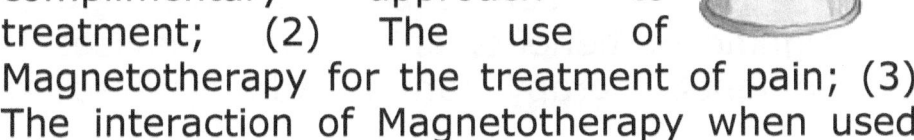

with other medical approaches; and (4) The complimentary aspects of using Magnetotherapy.

The Use of Magnetotherapy as a Complimentary Approach to Treatment

The history of magnets for the treatment of illness dates back through the history of several civilizations. Egyptian priests were known to use magnets to treat various conditions. In the fourth century Hippocrates documented the use of magnets for specific medical conditions. In the early fifteenth century a Swiss physician name Paracelsus 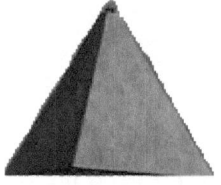 hypothesized that magnets had the ability to attract diseases out of the body (Santwani, 2005).

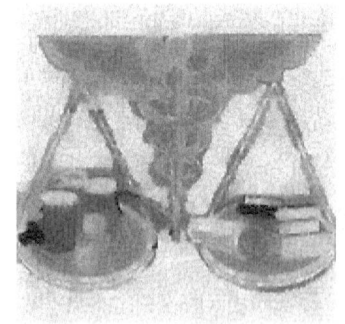

Today, there are many types, sizes and strengths of magnets being used as a compliment to both allopathic and naturopathic medicine. *What makes Magnetotherapy viable for use as a treatment?* One of the significant benefits of magnetic devices is that magnet therapy can be administered by patients independently, or by health care providers. Generally, magnetic devices are  applied to the whole body when chronic conditions are present, or to specific areas of the body affected by illness. There are a number of complimentary forms of magnetic devices which can be implanted or used externally. Modern medical applications using Magnetotherapy include devices designed to provide low frequency magnetic fields or those designed to deliver pulsed electromagnetic field therapy. 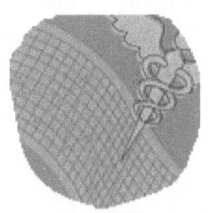 The applications use two specific types of magnets used to treat medical conditions: (1) Constant (static) magnets – these magnets contain a permanent magnetic field once they have been charged; and (2) Electromagnetic (temporary) magnets – these devices generate a magnetic field only while they are activated using electrical current.

 Other complimentary forms of magnetic devices include: (a) self-adhesive strips, (b) belts, (c) magnetic jewelry, (d) shoe and belt inserts, and (e) mattress pads. Modern complimentary methods of Magnetotherapy have incorporated magnet-conditioned water – a system using magnetized drinking water to treat various diseases.

Magnetization process was carried out by pouring these two solutions through Magnetic funnel. Finally, other methods such as magnet wraps are Lodestones (medicinal magnetic rocks) are sold for most body parts (Magnetic Technologies, 2007).

The Use of Magnetotherapy for the Treatment of Pain

Numerous reports have surfaced over the past several decades about the remarkable results of magnetic devices for the  treatment of chronic pain. Supporters of Magnetotherapy point to these findings as direct proof of this systems validity as a practical medicine. Presently there are no less than sixty five different medical conditions which are claimed to be cured or managed when magnetic devices are used for their treatment.

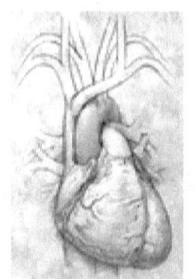 *How does Magnetotherapy influence pain?* Some practitioners have theorized that magnetic devices (1) influence blood circulation - increasing blood oxygen as it moves through the body, (2) alkalize bodily fluids – reducing the overall body acidity, (3) detoxify the blood vessel walls - removing plaques and cholesterol,

and (4) influence the levels of calcium moving through the blood. Any one of these specific systemic conditions affects pain within the body. Some of the more common uses of Magnetotherapy have been used to treat the following: (a) skeletal conditions such as bone fractures and osteoarthritis; (b) conditions of the central nervous system such as neurological disorders and nerve regeneration; (c)  muscular conditions such as multiple sclerosis and fibromyalgia; and (d) conditions resulting in generalized pain in the body.

 There is great need for more trials and testing to assess the overall effectiveness of magnetic healing. Skeptics continue to point the fact that a lack of scientific evidence renders any claims of effectiveness nothing more than wishful thinking. A recent report by the National Center for Complimentary and Alternative Medicine pointed to four specific practical tips for consumers: (1) use magnetic devices only under the direct supervision of medical professionals; (2) use magnetic devices for short periods of time; (3) report any side effects immediately; and (4) the need for consumers to stay informed when considering

the use of magnets therapeutically (NCCAM, 2004).

The Interaction of Magnetotherapy when used with other Medical Approaches

Some of the perceived benefits of Magnetotherapy include: (1) the ability of magnetic devices to change how cells function; (2) the ability of magnetic devices  to alter or change the balance between cell longevity and regeneration; (3) the ability of magnetic devices to act as a conductor in the blood, influencing the flow of diseased cells from the circulatory system; (4) the ability of magnetic devices to influence pain signals in the body; and (5) the ability of magnetic devices to influence bodily tissues on a cellular level. (NCCAM, 2004).

 Remarkably, there are relatively few specific negative indications when magnetic devices are used as a compliment to other therapies. It is noteworthy however to mention that magnetic devices should be used only it is clearly understood that their use presents no harm. With such a large number of conditions that have met with limited results

regarding the use of conventional pharmaceutical, many people have turned to Magnetotherapy simply because they lack other viable options.

The Complimentary Aspects of Using Magnetotherapy

 There are a variety of reasons why someone might choose Magnetotherapy as a compliment to other therapeutic treatments. First of all, in a general sense Magnetotherapy has been found to be safe and free of harmful side effects when used alone or with other treatments. Next, Magnetotherapy offers patients a choice in the type of treatment they receive. Many times patients feel a general loss of control when they are informed that they must use a particular treatment regime that has specific negative health indications.

In the end it is up to the individual user to determine the exact reason for using magnetic devices as a compliment to other therapeutic treatments. While Magnetotherapy has been used for a number of years there remains a cloud of doubt regarding its effectiveness as a treatment method. Practitioners and students should be aware of the potential criticism that will surface over the use of Magnetotherapy as a healing approach to medicine.

Chapter 6 – Understanding the Healing Processes of Magnetotherapy

"Using magnets to cure disease depends on its nature and symptoms"

Karen Menezes

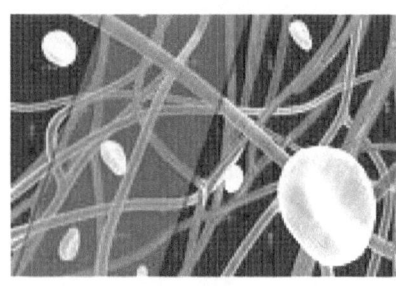

One of the most significant claims of Magnetotherapy is in its ability to affect the blood. The medical intervention of physicians working with the circulatory system is called hematology (Reference.com, *Hematology*, 2007). This chapter will address the use of Magnetotherapy from a haematological (circulatory) position. Several questions which will be addressed include: (1) Does Magnetotherapy truly affect the blood? (2) How does Magnetotherapy affect healing through the blood? (3) Can Magnetotherapy influence the properties of the circulatory system as a way to promote healing?

Does Magnetotherapy Truly Affect the Blood?

Without question the blood represents the most important fluid in the human body

(Santwani, pg. 22, 2005). With a volume of approximately five liters (one and a half gallons), and a major supplier of nutrients and minerals throughout the entire body, the blood stands out as the single most important substances connected to health in the human body.

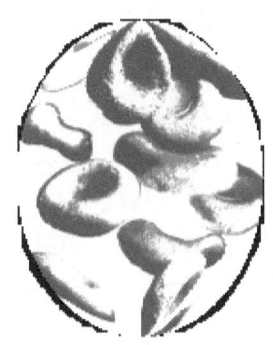 The fluid mix of blood includes: (1) red and white blood cells; (2) mineral salts; and plasma which serves as the main liquid compound. One of the principle functions of blood in the body is to carry oxygen from the lungs to every portion of the body, including organs and tissues. Blood has coagulants which are resident and necessary for the proper clotting effect in the event of injury to body tissues.

The flow of blood throughout the body takes place when the heart expands and contracts causing a pumping motion. This pumping motion sends the blood in a rhythmical motion through the arteries to the lungs, 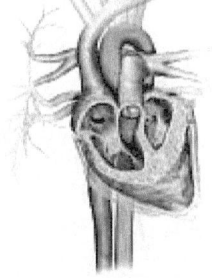 then to the major organs, and then to the secondary organs and tissues. When the oxygen has been furnished to a given portion of the body, the depleted blood returns to the heart so the cycle can begin again.

The whole motion of blood circulation takes place approximately seventy two time per minute. It is the electrical impulses of the heart that cause the process to take place. Because of the electrical activity taking place in the pulmonary system, there is an amount of magnetic energy that is traceable. This energy can be measured using magnetic devices such as an MRI (Magnetic Resonance Imaging).

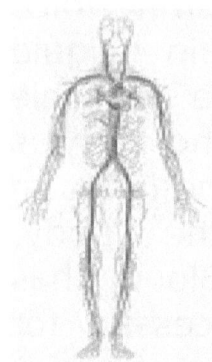

Supporters of Magnetotherapy theorize that it is possible to track the movement of energy through the circulatory system with magnetic devices. With the rapid flow of blood in the circulatory system it would be relatively easy to track specific conditions, assessing their movement and development. Theorists reason that it is possible to affect the blood using magnetic fields that can in fact help in the removal of specific toxins present in the blood system as the result of illness and disease.

When magnetic energy is applied, it creates ionization with the minerals present in the blood. This prevents clotting, allowing the blood to flow more freely through the arteries and veins (IBAM, *Magnetotherapy*, pg. 9, c. 2000). With more spontaneous flow, people with specific pulmonary difficulties have a better chance of surviving life threatening conditions such as heart attack and stroke.

There is great promise in the study of magnetic fields on the circulatory system. Because much is known about the presence and effect of hemoglobin on the blood, the results of tests and trials using magnetic devices has 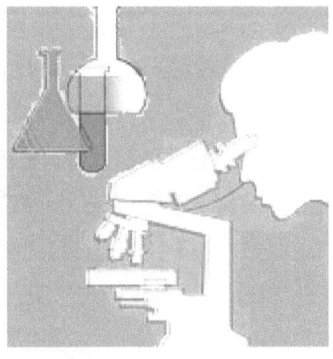 been promising. However, there remains many unanswered questions concerning the overall effect of magnetism in terms of health and physiology. Concerning the question of Magnetotherapy affecting the blood, it is reasonable to suggest that some effect does in fact take place when magnetic fields are applied.

How does Magnetotherapy Affect Healing through the Blood?

 The hematological effects of Magnetotherapy on the blood are significant in a number of ways. First of all, magnetism affects the overall flow of blood in the arteries and veins (Santwani, pg. 68, 2005). Tests have shown how the influence of magnetic fields affects the blood's fluid viscosity, conductivity, density, and mineral content. This fact is important, because

scientific medicine has discovered a number of direct connections between the make up of the blood and the ability of bodily organs to function optimally (IBAM, *Guide to Alternative Medicine,* c. 1998).

Secondly, a number of tests designed to measure the level of influence magnetic fields have on the blood have shown remarkable results in terms of separating certain elements from the plasma and other blood cells. This is an important 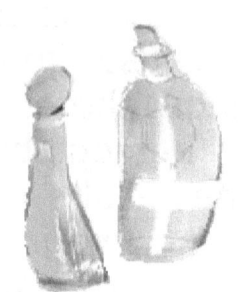 point in terms of identifying a specific strain of virus that can be researched when looking for cures for disease.

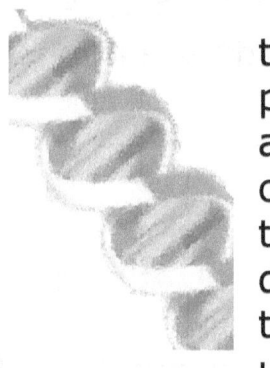 Finally, the magnetic influence taking place in the blood makes it possible to follow the direction of a given condition from organ to organ on a molecular level. Often, the elements of an illness or disease are left untraceable due to their relatively small size. By using magnetic resonance as well as other methods using magnetic fields researchers can identify specific conditions that would normally be passed over during a given procedure. The study of magnetism and the blood needs much more study and research in order to properly assess all of the associated benefits. However, more evidence is now

available concerning the activity taking place in the circulatory system.

Can Magnetotherapy Influence the Properties of the Circulatory System as a Way to Promote Healing?

This question remains one that leads to more questions rather than providing concrete answers. What is known is that there is a direct connection between magnetism and the blood. During the past several  decades scientists and medical researchers have made great strides in proving the viability of magnetic devices being used to treat medical conditions.

 As a system of therapeutic use, Magnetotherapy offers more promise than many of the skeptics are willing to admit. It has been the tireless efforts of scientists and medical researchers which have led to breakthroughs in the area of Magnetotherapy. Scientists have proved several things: (1) Cells in the human body have a magnetic field; (2) Magnetic fields produce electric currents in the minerals present in the blood as well as other bodily fluids; (3) When damaged cells receive magnetic energy they tend to regenerate more effectively with marked functional

improvement; and (4) The ionization of minerals in the blood has a direct correlation to the healing processes within the body. As a field of medical research Magnetotherapy holds great promise as a therapeutic treatment option for individuals who have seen little or no results from other methods of healing.

Chapter 7 – Procedures and Precautions for Using Magnetotherapy

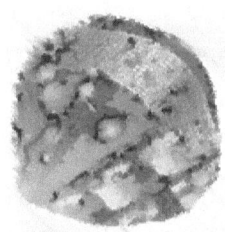

"Living systems are very sensitive to magnetic fields existing in the body"
Santwani, M.T.

The final chapter looks at some of the significant considerations regarding the use of Magnetotherapy as a therapeutic approach to healing. While it would be easy to say this therapeutic approach to medicine is completely free of medical concerns, the fact is that Magnetotherapy is a therapeutic system that should be used with caution.

The following considerations and precautions should be observed: (1) The use of magnetic devices is not the same for everyone. The effects of magnetism on the body varies from person to person. For this reason, anyone considering the use of magnetic devices for treatment of medical conditions should seek the advice of professional trained to work with them (IBAM, *Magnetotherapy*, pg. 13, c. 2000).

(2) Consumers and practitioners should be very aware of the cost factor connected to the use of magnetic devices used for the treatment of medical conditions. Magnet therapy has become a popular area for business minded people interested in earning money (Answers.com, *Magnetic Therapy*, 2007).

(3) People using magnetic devices should be aware of their specific use and indication on the body. In some cases, certain magnets could create an adverse effect on the body if the right device was not used (IBAM, *Guide to Alternative Medicine*, pg. 82, c.1998). The following points are helpful in deciding which device to use: (a) Know what condition the magnetic device is being used for; (b) Be aware of the magnetic polarity being used for a given condition; (c) Become aware of the bodies interaction with a given magnet strength and field; and (d) Always ask questions about the proper use of a device before using it for therapeutic purposes.

(4) The general precautions of use includes: (a) Discomfort or irritation caused by a magnetic device; (b) Specific circulatory and heart conditions that would be increased by the use of magnetic devices; and (c) Women who are pregnant. These are just a few of the common concerns associated with Magnetotherapy. Individuals interested in using magnetic devices should always consult medical professionals before beginning any type of Magnetotherapy.

Looking Ahead

It is likely that many more studies will be done regarding the use of Magnetotherapy. Readers are encouraged to continue their own research and study into the application of magnetic devices for the treatment of medically related conditions. As a field of natural medicine, Magnetotherapy certainly offers another option for people who are seeking help with their personal health and well-being.

References by Chapter

Chapter 1 – Magnetotherapy Defined

Barrett, Stephen, *Magnet Therapy: A Skeptical View*, (2002) Website: http://www.quackwatch.org/04ConsumerEducation/QA/magnet.html Retrieved on 2007-02-20

Carroll, Robert Todd, *Magnet Therapy*, (2006) Website: http://www.skepdic.com/magnetic.html Retrieved on 2007-02-20

Davidson, Michael W., *James Clerk Maxwell*, (2003) Website: http://micro.magnet.fsu.edu/optics/timeline/people/maxwell.html Retrieved on 2007-02-21

Dictionary.com, *Placebo*, (2007) Website: http://dictionary.reference.com/browse/placebo Retrieved on 2007-02-20

Gill, Lillian J., *Letter to William Roper*, (1997) Website: http://www.fda.gov/foi/warning_letters/d1153b.pdf Retrieved on 2007-02-20

IBAM, *Guide to Alternative Medicine*, (c. 1998) Calcutta, India, Pgs. 78-83

IBAM, *Magnetotherapy*, (c. 2000) Calcutta, India

IndianGyan.com, *Magnetotherapy and Naturopathy*, (2000) Website: http://www.indiangyan.com/books/magnetbooks/Magnto_Therapy/magnetotherapy_and_naturopathy.shtml Retrieved on 2007-02-21

Livingston, James D., *Magnetic Therapy: Plausible Attraction?* (1998) Website: http://www.csicop.org/si/9807/magnet.html Retrieved on 2007-02-20

MedicineNet.com, *Definition of Magnet Therapy*, (2007) Website: http://www.medterms.com/script/main/art.asp?articlek ey=22961 Retrieved on 2007-02-21

McCrory, P., *The Power of Placebo*, (2005) From: The British Journal of Sports Medicine, Vol. 39, Page 125, March; Website: http://bjsm.bmj.com/cgi/content/full/39/3/125 Retrieved on 2007-02-20

Ramey, David W., *Magnetic & Electromagnetic Therapy*, (1998) Website: http://skeptically.org/quackery/id4.html Retrieved on 2007-02-20

Santwani, M.T., *The Art of Magnetic Healing*, (2005) B. Jain Publisher LTD., Delhi, India

Schnack, Jurgen, *Quantum Theory of Molecular Magnetism*, (2005) Website: http://arxiv.org/PS_cache/cond-mat/pdf/0501/0501625.pdf Retrieved on 2007-02-21

Chapter 2 – The Theory of Magnetotherapy

Bellis, Mary, *Michael Faraday*, (2007) Website: http://inventors.about.com/library/inventors/blfaraday. htm Retrieved on 2007-02-26

Bonlie, Dean R., *Bio-Magnetic Theory*, (1999) Website: http://www.shokos.com/science.htm Retrieved on 2007-02-21

Doshi, K.M., *Structure of Sub-Atomic Particles: Magnetic Ring Theory*, (2001) Website: http://www.geocities.com/magneticringtheory/ Retrieved on 2007-02-21

IBAM, *Guide to Alternative Medicine*, (c. 1998) Calcutta, India, Pgs. 78-83

IBAM, *Magnetotherapy*, (c. 2000) Calcutta, India

IndianGyan.com, *What is a Magnet?* (2000) Website: http://www.indiangyan.com/books/magnetbooks/magn etic_cure/what_is_magnet.shtml Retrieved on 2007-02-22

Integrated Publishing, Inc., *NEETS: Module I – Introduction to Matter, Energy and Direct Current*, (2003) Website: http://www.tpub.com/neets/book1/chapter1/1h.htm Retrieved on 2007-02-21

Lee, E.W., *Magnetism: An Introductory Survey*, (1970) From: Dover Publications, Inc.: Website: http://www.rare-earth-magnets.com/magnet_university/history_of_magnetism .htm Retrieved on 2007-02-24

Meyer, Hans R., *Magnetic Theory*, (2005) Website: http://www.coilwinder.com/Theory%20of%20Magnetis m.htm Retrieved on 2007-02-21

Nave, Rob, *HyperPhysics*, (2005) Website: http://hyperphysics.phy-astr.gsu.edu/hbase/hframe.html Retrieved on 2007-02-26

Null, Gary, *Biomagnetic Healing*, (2007) Website: http://www.garynull.com/Documents/magnets.htm Retrieved on 2007-02-21

Santwani, M.T., *The Art of Magnetic Healing*, (2005) B. Jain Publisher LTD., Delhi, India

The Nobel Foundation, *Pierre Curie*, (2007) From: Nobel Web AB; Website: http://nobelprize.org/nobel_prizes/physics/laureates/1903/pierre-curie-bio.html Retrieved on 2007-02-26

Van Helden, Albert, *William Gilbert*, (1995) Website: http://galileo.rice.edu/sci/gilbert.html Retrieved on 2007-02-24

Vervliet, Ria, *Magnet Therapy: What is it?* (2005) Website: http://health.learninginfo.org/alternative-health/magnet-therapy.htm Retrieved on 2007-02-21

Wikipedia.com, *Electromagnetism*, (2007) Website: http://en.wikipedia.org/wiki/Electromagnetism Retrieved on 2007-02-23

Wikipedia.com, *Magnetism*, (2007) Website: http://en.wikipedia.org/wiki/Magnetism Retrieved on 2007-02-24

Wikipedia.com, *Peter de Maricourt*, (2007) Website: http://en.wikipedia.org/wiki/Peter_of_Maricourt#About_Epistola_de_magnete Retrieved on 2007-02-23

Wikipedia.com, *William Gilbert*, (2007) Website: http://en.wikipedia.org/wiki/William_Gilbert Retrieved on 2007-02-25

Williams, Andy & Burns, Vicky, *History of Magnetism*, (2007) From: University of Birmingham, UK; Website: http://www.aacg.bham.ac.uk/magnetic_materials/histo ry.htm Retrieved on 2007-02-24

ZES Bros., Inc., *Principles of Magnetotherapy*, (2007) Website: http://www.magnetocure.com/index.php?typ=MRC&sho wid=5 Retrieved on 2007-02-27

Chapter 3 – The Purpose of Magnet Induction for Therapeutic Practices

Bunny, Walter, *Man Wearing Nylon Over Wool Ignites Carpet with Static Electricity*, (2005) Website: http://www.reasoner.org/archives/300 Retrieved on 2007-02-28

Dictionary.com, *Induction*, (2007) Website: http://dictionary.reference.com/browse/induction Retrieved on 2007-02-28

Henderson, Tom, *Lesson 2: Force and Its Representation*, (2004) Website: http://www.glenbrook.k12.il.us/gbssci/phys/Class/newtlaws/u2l2b.html Retrieved on 2007-02-28

IBAM, *Guide to Alternative Medicine*, (c. 1998) Calcutta, India, Pgs. 78-83

IBAM, *Magnetotherapy*, (c. 2000) Calcutta, India

Santwani, M.T., *The Art of Magnetic Healing*, (2005) B. Jain Publisher LTD., Delhi, India

SeniorHealth.com, *Magnet therapy: Does it Really Relieve Pain?* (2006) Website: http://seniorhealth.about.com/cs/altmedicine1/a/magnet_therapy.htm Retrieved on 2007-02-28

Wikipedia.com, *Magnetic Field*, (2007) Website: http://en.wikipedia.org/wiki/Magnetic_field Retrieved on 2007-02-28

Chapter 4 – Types and Uses of Therapeutic Magnets

EBESCO Publishing, Inc., *Magnet Therapy*, (2007) Website: http://healthlibrary.epnet.com/GetContent.aspx?token= e0498803-7f62-4563-8d47- 5fe33da65dd4&chunkiid=33778#Work Retrieved on 2007-03-02

IBAM, *Guide to Alternative Medicine*, (c. 1998) Calcutta, India, Pgs. 78-83

IBAM, *Magnetotherapy*, (c. 2000) Calcutta, India

Santwani, M.T., *The Art of Magnetic Healing*, (2005) B. Jain Publisher LTD., Delhi, India

Chapter 5 – The Integration of Magnetotherapy with Other Therapeutic Treatments

IBAM, *Guide to Alternative Medicine*, (c. 1998) Calcutta, India, Pgs. 78-83

IBAM, *Magnetotherapy*, (c. 2000) Calcutta, India

IndianGyan.com, *Magnetotherapy and Naturopathy*, (2000) Website:
http://www.indiangyan.com/books/magnetbooks/Magnto_Therapy/magnetotherapy_and_naturopathy.shtml Retrieved on 2007-03-05

Menezes, Karen, *Understanding Magnetotherapy*, (2000) Website:
http://members.tripod.com/k_menezes/mag1.htm Retrieved on 2007-03-05

Pawluk, William, *The Positive Benefits of BioMagnetics*, (1995) From: Baltimore Resource Journal, Vol. 9, Num. 2, Summer; Website:
http://www.consumerhealthreviews.com/articles/MagneticTherapy/PositiveBenefitsofBioMagnetics.htm Retrieved on 2007-03-05

Santwani, M.T., *The Art of Magnetic Healing*, (2005) B. Jain Publisher LTD., Delhi, India

Wiancko, Ken, *Magnetotherapy, It Can Help You*, (2002) From: Consumer Health Reviews; Website:
http://www.consumerhealthreviews.com/articles/MagneticTherapy/Magnetotherapy.htm Retrieved on 2007-03-05

Chapter 6 – Understanding the Healing Processes of Magnetotherapy

IBAM, *Guide to Alternative Medicine*, (c. 1998) Calcutta, India, Pgs. 78-83

IBAM, *Magnetotherapy*, (c. 2000) Calcutta, India

Magnetic Technologies, *Magnetic Water*, (2007) Website: http://www.magneticeast.com/eng/health/magneticWater.asp Retrieved on 2007-03-06

NCCAM, *Questions and Answers About Using Magnets to Treat Pain*, (2004) Website: http://nccam.nih.gov/health/magnet/magnet.htm#consumers Retrieved on 2007-03-06

Reference.com, *Hematology*, (2007) Website: http://www.reference.com/browse/wiki/Hematology Retrieved on 2007-03-08

Santwani, M.T., *The Art of Magnetic Healing*, (2005) B. Jain Publisher LTD., Delhi, India

Chapter 7 – Procedures and Precautions for Using Magnetotherapy

Answers.com, *Magnetic Therapy*, (2007) Website: http://www.answers.com/topic/magnetic-therapy Retrieved on 2007-03-09

Consumer Health Reviews, *Biomagnetics: The Science*, (2002) Website: http://www.answers.com/topic/magnetic-therapy Retrieved on 2007-03-09

IBAM, *Guide to Alternative Medicine*, (c. 1998) Calcutta, India, Pgs. 78-83

IBAM, *Magnetotherapy*, (c. 2000) Calcutta, India

Santwani, M.T., *The Art of Magnetic Healing*, (2005) B. Jain Publisher LTD., Delhi, India

www.ingramcontent.com/pod-product-compliance
Lightning Source LLC
Chambersburg PA
CBHW021922170526
45157CB00005B/2143